HOW WILD ANIMALS
Fight

HOW WILD ANIMALS Fight

By Dorothy E. Shuttlesworth

Doubleday & Company, Inc., Garden City, New York

Library of Congress Cataloging in Publication Data

Shuttlesworth, Dorothy Edwards, 1907–
How wild animals fight.

SUMMARY: A description and explanation of the various ways in which wild mammals fight and the reasons why they do, such as self-defense and protection of young.

1. Animal fighting—Juvenile literature. [1. Animal fighting. 2. Mammals—Habits and behavior] I. Title.

QL758.5.S58 599'05
ISBN 0-385-08596-6 Trade
ISBN 0-385-08599-0 Prebound
Library of Congress Catalog Card Number 75-19144

Printed in the United States of America
First Edition

For Rich,
who likes to settle problems peacefully

Contents

HOW WILD ANIMALS

Fight

1. *Survival*

SURVIVAL IS a problem for everything on earth. Even plants have a struggle to survive if the soil where they grow is changed in some way, or if weather conditions become unfavorable. But struggles for survival by plants do not seem so dramatic as those of the animals.

Wild animals of jungles, woods, fields, and plains meet a variety of dangers. The greatest of these are caused by people who take over their natural homelands for building or farming, and people who kill animals whose fur or other features can be sold for profit.

Animals are helpless against these enemies. There is no way to fight back. However, in their own wild world there is often combat.

When wild animals fight, the results sometimes are unexpected and tragic. Here is a situation where two white-tailed deer engaged in a duel, and both were losers. Their antlers locked in such a way that neither could free itself. (Wilford L. Miller from National Audubon Society)

Flesh eaters fight the enemy of hunger by attacking other animals. And some of those attacked are able to resist.

At times animals fight members of their own species. Often this is to establish the right to have certain territory for their own.

Male animals also battle to win a mate, or to claim the leadership of their group or herd. Females fight to protect their young. Sometimes males do this also.

Methods of fighting depend on a variety of weapons. There are horns and antlers, hoofs, teeth, claws, and quills. Bad odors and armor are aids that help in defense.

Struggles for survival take place not only among mammals on land, but among animals of the oceans, seashore, and inland waters, and among birds and reptiles. However, in *How Wild Animals Fight* we will look mostly at the mammals. Their behavior alone makes a great story for all who are interested in our natural world.

2. *Horns, Heads, and Antlers*

THE HORNS OF a rhinoceros are unusual. People usually guess they are made of bone, but this is not the case.

In fact, a rhino horn is made up of a substance that contains a tough protein called keratin. This grows out of the skin, as toenails and hair do, but it appears only over the nostrils. As it forms a wide base, then narrows to a sharp point, it creates the horn—a frightening weapon. This continues to grow throughout the animal's life.

The mighty rhinos, living in the warmer parts of Africa and Asia, have no need to fight for food. These animals eat only grasses, leaves, and twigs. However, they are well equipped to do battle. One of them can hold its own fighting even a great elephant, if these two happen to cross each other's path.

Actually, rhinos have violent tempers. When they have been frightened or startled in any way, they are likely to rush about, trampling bushes and cutting up the ground.

The strength and size of a rhinoceros, as well as its great horns, give this animal tremendous fighting power. (Mark Boulton from National Audubon Society)

Then if they see anything move, no matter how large it might be, they rush at it, aiming to give a terrific jab.

When a rhino charges like this, it may move as fast as thirty-five miles an hour. At the beginning of the charge its head is carried high. Then it is lowered for a killing thrust with the horns.

If a rhino calf does not stay close to its mother, it can easily become the prey of a great cat or even of a smaller hyena.

The hyena will cleverly tease a calf into moving away from its protective mother. As the young one gives chase, the hyena suddenly turns and tries to grab a leg, an ear—anything it can sink its teeth into.

A terrified scream from the calf usually brings the mother at top speed, with horns pointed toward the attacker. The hyena is then likely to release its hold and quickly run away.

Calves are defenseless against hyenas until they are about four months old. If one has lost its mother, it may be "adopted" and given protection by a powerful bull.

There are two different kinds of rhinoceroses in Africa. They are popularly called the "black" and the "white" although there is really not much difference in the color of their hides. The two species are somewhat different in size (the white are larger, weighing up to four tons) and there are some slight differences in their habits. But in both kinds, females as well as males wear two horns.

The great Indian rhinoceros of Asia grows only a single horn. However, it can be just as violent, if not more so, as its two-horned relatives.

A number of other animals have horns partly made up of the same kind of substance that forms a rhinoceros horn. But these horns have a core of bone.

When a mountain goat, for example, is a few months old, two small bony lumps develop at the front of its skull—one on either side—to form the core of the animal's horns.

From then on, cells in the skin around the base of each lump give out the substance that contains keratin, and for a few years the mountain goat's horns grow rapidly—slim, straight, and daggerlike. After that, the growth rate is slower.

Both males and females have horns, and both put them to good use when flesh-eating animals threaten them. As the goats graze over steep mountain slopes, eagles dive at them, and bobcats and mountain lions may attack at any time.

A female with a young one may try to cover the kid with her body when she is threatened. Then she jabs at the enemy with her horns. Often she succeeds in driving the attacker away.

Fully grown, mountain goats are big; they may weigh two hundred pounds or more. As adults they are not often bothered by hungry animals on the prowl. However, they are not always peaceful among their own kind.

During the mating season a male will stalk around a rival, with legs held stiff and head lowered. After this challenge, a fight with much bloodletting may develop, as each contender plunges his sharp horns into the chest, belly, or flank of his opponent.

These goats have a neighbor in their homelands—the Rocky Mountain bighorn sheep—that wears horns of a very different design. Though these horns are produced in the same way as those of the goats, the ends curl downward instead of growing straight.

Each year a ridge of new growth develops at the base, and the horn continues to grow—and curl. It may finally make a complete circle, and even continue to spiral beyond one full turn. The horns of full-grown males become enormous. Those of females are smaller.

As with the mountain goats, male sheep fight each other during the mating season. But the "bighorns" are not likely to spend time in challenging. Two rams, if interested in winning the same ewe, quickly back off from each other, then rush forward. Their speed is great, and when their mighty horns come together, the crash can be heard a mile or two away.

The first charge is followed by a second, and another and another, but the enormous curled horns serve only for battering, not for stabbing. The shock of their impact does the damage, and the ram with the most strength and endurance becomes the winner. The fighters do not battle "to the death."

On the other side of the world from these sheep and goats of North America, we find another powerful horned beast—the water buffalo of Asia.

The horns of this enormous animal are a very different style from those of goats or sheep. They are long, and go backward in a sweeping curve. Unless a buffalo is seen in action, it is hard to imagine how they can be useful weapons.

However, if we were to watch a buffalo do battle with a tiger, we would see that each motion of its head brings the horns' tips around closer to its neck. And that is the area the tiger is attempting to grab in its teeth.

A fight between these two is vicious. Either one may win. Or both may lose their lives in the battle.

The horns of a giraffe are quite useless for fighting. They are short, covered with hair, and padded at the tip. They could scarely pierce an animal's skin, let alone its flesh.

But the shortcomings of their horns do not prevent giraffe bulls from fighting. The entire head is used to strike blows! And because the head may weigh as much as a hundred pounds, its force is great.

In a fight against a rival, a giraffe swings his head and neck like a mighty hammer moving sideways. If it

These giraffes may seem to be pictured in an affectionate pose, but actually they are fighting. When a male tries to strike blows with his head, his opponent may protect his own head and neck by moving close to the attacker. (Leonard Lee Rue III from National Audubon Society)

hits its target, the neck of the opponent may be broken. The one that is trying to defend himself moves up close to his opponent, but in such a way as to be out of striking range.

Feet also prove good weapons for these animals. They can kick out with their long, powerful legs and deal a serious blow with the hoofs.

Except for occasional fighting with their own kind, giraffes seldom have reason to be anything but peaceful. Their one natural enemy is the lion, and even a lion hesitates to attack a full-grown giraffe. If, however, two or more of the big cats discover a giraffe with forelegs spread wide apart, reaching down to drink at a waterhole, they can easily make it their victim.

Speed is a giraffe's best defense. In open country, and even in the bush, one can race along at more than thirty miles an hour.

Deer and their close relatives use antlers when they fight. These branched headpieces often are confused with horns. Let us see in what ways the two are different.

Antlers are composed of bone; they do not grow out of the skin as horns do.

Antlers branch out, taking on many different shapes. Horns are not branched.

In most cases horns remain on an animal's head for life. (An exception is the pronghorn antelope, which sheds its horns annually.) Antlers grow for several months, then fall off as a new set begins to develop.

Antlers serve male deer only, with the interesting exceptions of caribou and reindeer. In the horn-bearing animals such as rhinos and goats, males and females carry the weapons.

Moose are the largest of all members of the deer family, and they grow antlers suitable for their great size. A bull may weigh a thousand pounds or more; his antlers may weigh fifty or sixty pounds.

Antlers have sharp spikes, or tines, but these are not always efficient in combat if they are not well placed or long enough to be effective However, the wide, flat palms are helpful in defensive fighting.

The fierce duels that take place between two males that are trying to win a mate usually are no more than noisy, blustery shoving matches. The bucks move toward each other, lower their heads, and suddenly lock antlers. Then, bracing their hind legs, they try to push each other backward. There may be five or six "rounds" in such a sparring match.

Usually the weaker of two fighters ends the struggle by running away. But sometimes wounds are suffered. And there is always the danger of a miserable death for both contestants if their antlers become locked. Then the animals lie helpless, unable to move, until they starve or are set upon by flesh-eating animals.

After a duel between two moose, the area where they have been fighting tells the story, for the ground is torn up and many trees are badly scarred. The antlers of the fighters also are scarred. Tines are broken off, and even the broad parts of the antlers may be battered.

Male moose fight each other to win a mate. The two embattled contestants, using their antlers, try to push each other backward until the weaker one gives up. (Leonard Lee Rue III from National Audubon Society)

Battered and broken antlers are not a long-time problem, however, because when the breeding season is over for the year, they are shed. By the following spring two knobs of soft bone develop on the buck's forehead. They are covered with a fuzzy skin, called velvet. Rapid growth brings them to their full growth in a few months. Then the bone hardens.

Soon the buck is rubbing his antlers against trees and rocks, scraping off the velvet, which has become dead tissue. Even after that has gone, the rubbing and scraping continues as the antlers are made sharp and bright—ready for active service.

Antlers become larger and more elaborate with each passing year until a buck reaches the prime of his life. Then for several more years, each new set remains about the same. After that, if he survives to old age, the antlers become smaller each year and are not so fine a quality.

Next in size to the moose in the deer family is the elk, and the widespread antlers of the bulls branch out into numerous tines. They are both handsome adornments and effective weapons.

Elk are famed for their battles during the mating season, and for the "bugle call" which males send out as a challenge. Taking a stand in his woodland home, a bull begins his call to battle in a low key, but gradually this rises to a high, wild pitch. Suddenly the sound drops into a lower key, with harsh screams and grunts added to the original tone.

Another bull, accepting the challenge, answers with a loud bark—and the stage is set for action.

When the two begin their combat, their necks swell and the nostrils flair. And they rush at each other until their antlers crash together. Sometimes the force of the crash is so great an animal has his neck broken. But more often the first charge is followed by another and another until one gives up and gallops away. A bull continues to challenge others until he has won a number of cows to his harem.

Caribou and reindeer are the most "antlered" of all deer. Females as well as males wear them. And the fawns begin to grow small spikes when only two months old.

The antlers sweep back gracefully over the neck, then bend forward, outward, and inward again. They are very handsome, but once the mating season begins and the males start to fight with each other, their antlers soon look rather shabby. The spikes are more delicate than they appear and are quite easily broken.

3. *Hoofs and Claws*

Hoofs and claws may not look like weapons; they may seem to be nothing more than part of an animal's foot. But if we see them in action, we know how deadly they can be.

Even deer often fight flesh-eating attackers with their hoofs instead of using their antlers. They rear up on their hind legs and kick out with the front, slashing out with the hoofs. Sometimes the hind legs and hoofs are used for the action.

Hoofs are especially useful as weapons during the time a deer is without antlers, or while new antlers are growing.

Female deer use their hoofs too. They use them in their own defense, and sometimes to defend their calves.

There comes a time when a mother may use her hoofs against her own young one. This happens after a calf has lived close to its mother for a year and the mother has a new baby to care for. Then she cannot give any attention to the yearling, yet the young one may not be

willing to go off on its own. So the mother strikes out with her hoofs in a manner that can't be mistaken—and the yearling soon retreats.

Once in a while a scrap starts between two females, and the opponents "square off," standing on their hind legs, and using their front hoofs as weapons.

The hoofs we are most apt to think about are those of horses, because horses are domesticated animals that we often see. But the horse you may ride at a farm or on a bridal path, or watch at a race track, has the same kind of hoofs as wild horses, zebra, or other members of the horse family.

Hoofs develop from the outer layer of skin, as horns do.

Thanks to fossils, we are able to follow the story of horses' hoofs very exactly. The earliest horse ancestor, called the "dawn horse," which lived about sixty million years ago, had neither hoofs, horns, or claws with which to protect itself. Its only hope of escaping flesh-eating attackers was to race away from them.

So the little dawn horse (it was no more than eleven inches high) and its descendants became swift runners. And through millions of years the center toes, which bore

When female deer fight among themselves, as they do occasionally, they stand erect and use their front hoofs as weapons. (Dade Rayfield from National Audubon Society)

most of the animal's weight, grew larger and larger, while the outer toes grew gradually smaller. At last all that was left was one very large toe encased in a horny covering.

Today this hoof structure of horses remains: one toe covered to form a hoof. With other animals that also have hoofs, such as rhinos and camels, more than one toe is encased in each hoof. But horses are the most successful in using their hoofs as weapons. They can kick out with either front or hind feet, and give an enemy serious trouble.

Kangaroos do not have hoofs, but their feet can be used as powerful weapons. These are tipped with strong nails that can rip and slash like so many knives.

Family fighting is carried on between an old male who is leader of a group and its younger members. When the old "boomer" is challenged by a young male, he takes strong action, beginning an attack by biting and boxing with his front feet. If his opponent weakens, he finishes the battle by leaping up and slashing him with his back claws.

An additional fighting aid is the tail. This is large, long, and strong, and it can support the kangaroo's whole body, giving the animal freedom to use all four feet in a desperate struggle.

Domestic horses, like their wild relatives, depend on hoofs for both playful and serious fighting. (Wide World Photos)

A kangaroo, though a peaceable animal for the most part, can fight savagely. The strong claws on all feet make knife-like weapons. And a tail strong enough to give support to the whole body, enables one of these animals to use all four feet at once. (Arthur W. Ambler from National Audubon Society)

But for the most part, kangaroos are peaceful animals. If threatened suddenly by wild dogs or other enemies, they try to escape by hopping and leaping out of reach. However, if one is cornered, it will try to defend itself. A large male, if he can get his back against a tree, can stand off a whole pack of dogs, slashing each one as it approaches him with the claws of his hind feet.

It is not surprising that such big animals as kangaroos can, and will, fight. But we hardly expect "timid" rabbits and hares to be able to defend themselves and sometimes actually start an attack. Yet we find this is a normal part of their lives.

If a snake threatens her babies, a mother cottontail will leap at the reptile, clawing and biting ferociously. If she is quick enough, she may overwhelm her enemy. Usually the snake tries to escape.

Rabbits and hares are preyed on by many mammals, such as wild cats and foxes. And large birds such as hawks and eagles depend on them for food. Usually they are helpless against these animals. However, sometimes they are able to "get the jump" on larger, stronger opponents.

The fur over a rabbit's toes covers long, sharp nails. These are effective weapons.

Fighting for a mate among the males takes place every spring. In the far north, the Arctic hares end the quiet life of winter, when they had only friendly relations with each other, to battle for does. They snap viciously at their rivals and, using claws and teeth, may rip chunks of hair out of their opponent's throat. Sometimes

they stand on their hind legs and box, kangaroo-fashion. Jack rabbits, living in more comfortable climates than the Arctic hares, behave much the same way. In fact, they are famous for their combats during the breeding season. Often males show ripped ears and other scars that have been inflicted during a fight.

European hares also are ferocious fighters. They leap high in the air as they attack each other with claws and teeth.

A raccoon has a bright, amusing face. In spite of the black fur across its face that suggests a bandit's mask, it looks more friendly than frightening. However, when it is threatened, this animal will put up a strong fight.

Raccoons do not have many natural enemies, but they sometimes tempt flesh-eating mammals. The cubs especially suggest "good eating." So, a coon mother, walking through the woods with her youngsters, is always alert for danger.

If she senses an enemy near by, she quickly pushes the cubs up a tree trunk, then runs swiftly through the underbrush, swamps, or whatever lies ahead. The en-

Hares and rabbits, supposedly timid, can fight bravely. This hare, feeling threatened by a raccoon, attacked fiercely, and it was pictured as it "got the jump" on its opponent. (Lynwood M. Chace from National Audubon Society)

emy—if there really is one—may follow, or may give up the chase. If it does catch up with her, there is a furious fight, and the coon may well be the winner. She then returns to the tree where her cubs are waiting.

Often raccoon enemies are not wild animals because in America coon hunting with dogs has been a popular activity for many years. Hounds and other breeds of dogs are raised especially to be coon hunters, and it is said the extraordinary fighting ability of raccoons is the reason some people enjoy this unhappy "sport."

A coon chased up into a tree sometimes plunges down on a dog's back, knocking it over and then escaping in the confusion that results. If it tackles a dog, it does so with vigor. It lunges at its opponent with teeth bared. Then it may run backward (which it can do almost as swiftly as running forward) only to suddenly rush forward again, snarling and lashing its tail from side to side.

When the fighting gets really close, the coon rolls over on its back and, gripping the opponent with its forelegs, tears with its teeth and rips with the claws of its hind feet.

Raccoons have long, sharp claws which are always exposed, ready for action. By contrast, claws of members of the cat family are drawn into the paws when not being used. However, this does not make them any less useful. Once they are imbedded in a victim's flesh, any struggle makes them sink in to a greater depth.

Dogs that are trained to be raccoon hunters find their prey to be clever when pursued and fierce if caught. The raccoon may roll on its back, grip the dog with its forelegs, and then use its teeth and claws. (Leonard Lee Rue III from National Audubon Society)

Do great, powerful bears have natural enemies? It would seem they are too large and frightening for flesh eaters to attack.

For the most part this is true. But a pack of wolves may, on occasion, overcome a black bear, or a hungry cougar might try to get the best of one. These happenings are unusual, however.

Sometimes there are fights between bears and moose or elk. Usually the bear starts the trouble, and the antlered animal turns against it in self-defense. The antlers may be used, but usually a few sharp blows with the hoofs defeat the bear.

Grizzly bears are among the giants of the animal world. They usually weigh about five hundred pounds, and sometimes are much heavier. With a single blow of its enormous paw, a grizzly can crush the skull or break the back of almost any creature.

Besides its size and strength, the grizzly's paw has long, heavy claws—larger than those of any other kind of bear. But they are blunt rather than sharp, and it is believed they are used not so much as weapons as for digging into the earth for food and for catching slippery fish.

The muscular front legs, huge paws, and enormous claws of a grizzly bear give this giant deadly striking power. (Leonard Lee Rue III from National Audubon Society)

Grizzlies have a reputation for being extremely vicious. However, studies of their actions seem to prove that much of this reputation grew because of the way they were hunted.

Since the early days of exploration in the United States, these bears have been shot down, and it could be that the smell of gun powder and the scent of men became as one to them. Not surprisingly, then, this scent became identified as "enemy."

The grizzly's evil reputation grew, also, because of a custom in Old California when the bears were used for bull fighting. One especially large bear would be put into an arena and a maddened bull would be turned loose to charge at it. Then the spectators would be thrilled if the grizzly managed to rise up on its hind legs and strike the bull dead with a well-aimed blow.

People, without guns, observing wild animals in their natural surroundings, have watched grizzlies come in close contact with other animals—even mountain lions—and saw that no fighting resulted.

It is true that bear cubs "play rough," and sometimes seem to be fighting as they wrestle, punch, spar, bite, and tumble each other. But mostly this is in fun. Often the young ones play this way with their mother who usually seems to enjoy their roughness.

Fights between older bears can be serious and deadly, however, as when one male fights another to keep him away from a chosen female. Also a male may attack a small cub, and if the cub's mother is near by, the attack can quickly turn into a battle between the two adults. They may stand up to each other, dealing blows to the opponent's head and shoulders; they may fight on all fours as they try to bite the opponent's jaws and throat.

Rival male bears fight each other in much the same way. If one is thrown to the ground, it usually gives up and escapes as quickly as possible. But some fights are to the death.

During a battle the bears are likely to bellow with rage. This is a frightening sound that fills the surrounding forest.

4. *Daggerlike Teeth*

THE TEETH OF an animal reveal a number of facts about its life story. They indicate what type of food it eats—whether plants or flesh—and therefore show if it must kill in order to survive.

We can learn from fossil teeth about animals that died away long before people lived on earth. Looking at a skeleton of the dinosaur *Tyrannosaurus,* we see its sharp-pointed canine teeth, and they show that this was a killer. In contrast, the flat-topped teeth of *Brontosaurus* certainly could not have served to bite either in self-defense or for attacking. *Brontosaurus* browsed in swamps, eating plants—and trying to keep away from the flesh eaters.

Animals living today have the same variety in teeth. And, as with creatures of the past, their food must be influenced by their teeth.

If we have a close-up look at the mouth of a tiger, or any other flesh-eating animal, we see long, sharp canine teeth. There are two in the upper jaw and two in the

Tigers and other great cats have the sharpest and longest canine teeth—in proportion to the size of the animal—of all mammals. These teeth are tremendous weapons both for killing prey and fighting rivals. (Franklin Williamson from National Audubon Society)

lower, placed on either side of the incisors, or front teeth.

The incisors of plant eaters, such as horses, are shovel-shaped, but sharp enough to cut leaves and grasses. With rodents, the incisors are extremely sharp—sharp enough to gnaw through wood. Beaver can cut down trees with their teeth.

Besides all the ways in which teeth are used in the everyday business of living, most animals use them also for fighting. It is instinctive to defend themselves, or to attack, by biting.

When we think about "frightening" teeth, the type most likely to come to mind are those of the great cats. Tigers, leopards, and their close relatives have the sharpest and longest canine teeth (in proportion to the size of the animal) of any of the mammals. Cats also have the sharpest claws of all the mammals.

Among these great cats, leopards are rated the most threatening to other creatures. This is because a leopard can outrun and out-jump others and is a strong swimmer as well. It can easily leap ten feet or more in the air and run up a tree trunk with amazing speed.

Besides having these physical abilities, a leopard is smart; it remembers anything helpful to its survival. It learns that a pack of baboons can fight back and are best left alone, but a single baboon is usually an easy prey. The leopard leaps for its neck and, if possible, immediately sinks its teeth into the jugular vein.

A leopard is clever also in its approach to a victim. It may lie very quiet, waiting for the right moment to spring. Or it may roll on the ground and make itself noticed by unusual playful motions. Seeing this from a distance, a deer will become curious and move closer. The leopard continues its play-acting until the deer is only a quick jump away—and at that point the deer is doomed.

Tigers and lions use their teeth much as a leopard does. However, they use somewhat different methods of attack. Lions often lie in wait near a waterhole. Then when a thirsty animal comes to drink, the big cat creeps toward it, growling. When fairly close, the lion gives a fast spurt, perhaps going at a rate of more than forty miles an hour.

Sometimes a lion gallops alongside a zebra, then suddenly slaps it on the neck with a huge paw.

But lions are known as peaceable animals; they hunt only when they need food. Among themselves they are friendly and sociable. Usually they do not fight over prey; several will share one carcass without any ill will. And they do not have to be on guard against natural enemies for other animals are aware of their power.

Fights do occur during mating seasons when a male may battle two or three rivals to keep them away from his chosen mate. Such fights involve ferocious snarling and the use of powerful paws and teeth.

Smaller animals, even if they have daggerlike teeth, find different means of hunting and fighting.

In Asia the so-called wild dogs, or dholes, hunt in packs. These animals are only about three feet in length and have rather short legs. However, when a dozen or more close in

Wild dogs have sharp canine teeth, but because the animals themselves are not very large, they work in packs to bring down prey such as deer. Sometimes they succeed, sometimes they lose out. (Courtesy of The American Museum of Natural History)

on a large beast such as a tiger or deer, a prolonged battle usually takes place.

The wild dogs surround their prey and take turns rushing at it, their teeth bared, ready for a quick bite. Gradually the larger animal grows tired, and may be overcome by the pack. However, each time a dog comes close, its life may be ended as it is crushed by a deer's hoof, or its back broken by a blow from a tiger's paw.

Either side may win or both may lose, for although the larger animal may be killed, it may first fatally injure all its attackers.

When a tiger attacks a large animal, its teeth are dependable weapons. It is likely to leap on a victim's back, then sink its sharp canines into the neck.

Wolves, like wild dogs, are likely to hunt in packs. Working as a team they can take on such opponents as a deer, which may weigh more than twice as much as any one of them, or a moose that weighs many times as much.

If a moose refuses to be panicked by the snapping wolves, it usually is the winner. The smaller animals try to surround their prey, worrying it until it tries to run away. Then they can close in from the rear, biting at its legs and hind quarters. However, if the moose does not run, but stands its ground for about five minutes, the wolves usually give up. They move along to look for a victim that is easier to take.

Years ago some people who wrote about wild life stated that wolf packs carefully organized their hunts.

Plans were supposed to be carefully worked out, with each member knowing what moves to make. But more recent studies of wolf behavior indicate that when a pack sets out to make a kill, each individual acts on its own. However, parents do teach their young and, to some extent, guide them in hunts.

Unlike large members of the cat family, wolves are *running* animals. They are not built for stalking and pouncing as cats are.

Spirited fighting sometimes takes place between wolves and bears, as when a bear prowls around a wolf den, looking for food that the wolves have dropped or buried near the opening. When the den's owner discovers the prowler, the bear may quickly retreat. But if not, the fight begins.

A bear has greater size and weight than a wolf, and a protective covering of long, thick hair, as well as strong paws and teeth. The wolf depends on teeth only, but it moves more easily and quickly than a bear, and the two may be quite evenly matched. Of course if only one bear is discovered by several wolves, or if several bears go food-hunting together, the fight can be very one-sided.

A number of accounts have been given of "running fights" where several wolves chase a bear for at least ten minutes, snapping and snarling viciously. However, they are careful not to get close enough to be struck by the bear's paws, and after the giant is well away from their den, they let it escape.

A wolf pack is usually made up of a male and female

and their latest litter of pups. There may be uncles and aunts, and perhaps others that are not close relatives. The average size of a pack is ten individuals.

Within a pack there is a "social scale." In a small pack this may cause few problems, but in a larger pack things are different if several males try to take the leadership. This naturally causes in-fighting until the question of which is the strongest is settled. Afterward the pack peacefully carries out its routines of hunting, eating, and sleeping.

When a leader begins to approach old age, which he does after about ten years, his place is taken by a younger, stronger member of the pack. Three or four years later the teeth of the aging wolf are so badly worn he cannot fight, even for food. Then he is lucky to find animals already dead for his meat supply, and he will eat berries, fruit, and almost anything that can be swallowed.

As with many animals, wolf groups have a certain area of land which they feel is their own "home territory." The territory for a family may have a diameter of nine miles or more.

Within this area members of the family have their dens, raise their young, and hunt. They do not object to other kinds of animals roaming through it, but wolves from the "outside" are not welcome, and are usually attacked if they try to enter.

Like wolves, a fox family has its own special territory, and in the far north, where food is especially scarce, problems arise. Parents take good care of their babies, feeding and protecting them. But when the cubs, born

Sometimes a dispute over territory causes fights between rival foxes. Competition is especially keen in areas where food is scarce. (Leonard Lee Rue III from National Audubon Society)

in the spring, have had a year of this attention, they must leave their parents' territory to find another not already settled by a fox family.

Often such places do not exist. Then the young ones either become the victims of flesh-eating animals or die of starvation. Even then, of course, they serve as food for prowling, hungry creatures.

In warmer climates foxes have somewhat different problems, for there are a greater variety of enemies. Besides wolves, there are bobcats, mountain lions, and lynx. And, more than in the far north, there are people—people who not only trap them for their fur but—as with raccoons—hunt them for what some consider to be a sport.

Foxes, pursued by hunters and dogs, are famed for their skill in outwitting these enemies. As they run they may backtrack for a distance, then jump sideways and start off in another direction. Or they leap on a stone wall or run through a shallow stream so that their scent disappears. Such tricks usually are successful for a while in confusing their pursuers and may give the unhappy fox a chance to hide.

Some of the same cleverness is shown against natural enemies. When the father of young cubs is guarding his den and a predator such as a lynx appears, he tries to draw attention away from his home by running.

If the prowler chases him, he may not return to where the cubs are hidden. However, if this strategy fails, the fox, using his teeth as weapons, fights bravely to defend his family.

We usually think of seals and sea lions as being water animals. They roam through the oceans as they hunt food, eat, and sleep. Fighting each other, or any other kind of animals, does not seem likely there.

But these mammals belong to the land as well as to water. Though they swim so expertly and have flippers instead of feet, their breeding grounds are on rocky islands and coast lines.

There is interesting variety among seals. The "leopard" is one kind, and it is noted for having the most fearsome teeth of all the species. Those in front are long and sharp, with big canines at either side. The jaws are the largest on any kind of seal.

Leopard seals are the only species that attack and eat warm-blooded animals. They also go after any other creature that falls into the water or lands on it. Penguins often become their victims, and sometimes other seals do also.

A more usual diet for seals is fish. Because seals' teeth point backward, they are well suited for grabbing their slippery prey and forcing it back toward the throat.

Largest of all is the elephant seal. Its name is suitable because of its huge size and because of its long nose which suggests an elephant's trunk. In spite of their great bulk (a male may weigh more than five thousand pounds) elephant seals move gracefully in the water, and they go far into its depths pursuing fish, including small sharks.

When fighting between two elephant seals begins, the opponents face each other with their heads held high. Then their teeth are bared and the attack starts in earnest. (Leonard Lee Rue III from National Audubon Society)

When they come on shore for the breeding season, there is great commotion. As with other seals, each bull tries to stake out territory which will be his alone. And as the females join them, each fights to win a number of them for himself.

A battle begins with two bulls facing each other, roaring, then attacking, with teeth bared and head held high. Each makes frightening sounds, with its elephant-like nose swelling mightily from the air blown into it.

Fur seals are said to be the most vicious fighters in this group of mammals. Battles take place when one bull tries to invade another's territory. As the defending bull charges at the invader, he may jump up on all four flippers and practically fly at his rival before seizing him by the neck and shaking him violently.

If we were watching, we would think this was to be a fight to the death. But after a few moments the two contestants pause to roar and hiss at each other. Then, usually, the invader hurries away as fast as his flippers will take him. The defending bull does not follow to seek revenge. He is satisfied to have his rival leave.

5. *The Longest Teeth*

THE CANINE TEETH of wolves and tigers are concealed weapons—not seen until they suddenly are bared. There are other teeth that are so long they cannot possibly be hidden inside a mouth. We call them tusks.

What would a walrus be without its tusks! Here is an animal living among floating ice in the freezing waters of the far north. With its clumsy body it cannot move as quickly on the ice as a polar bear, nor be as swift in the water as a killer whale. And both of these animals are ever anxious to make a meal of a young walrus. Tusks are its means of defense.

When a walrus mother realizes that her young one is being stalked by a polar bear, she rises to her full height, gives a great bellow of defiance, and moves closer to her calf.

If the bear is not frightened away, the walrus attacks, bringing her tusks down on the enemy with all her might. The sharp points can wound the bear badly, and it may crawl away to lick those wounds. Or it may

Tusks are extraordinary teeth that make it possible for a walrus to attack such foes as polar bears and killer whales. (Leonard Lee Rue III from National Audubon Society)

finally manage to outwit the mother and snatch the young one.

In the water, the fight against killer whales is more uneven. Again the walrus uses its tusks but as soon as it sinks its great teeth into one of these "tigers of the sea," as killer whales have been called, another and another attacks, for these killers travel in schools.

When a young calf falls between two chunks of ice and becomes wedged there, the mother slashes at the ice as if it were indeed an enemy until her young one is freed.

Among walrus herds the size of an animal's tusks has much to do with which are the most important individuals. Of course, the larger the tusks, the greater the importance.

Mostly walruses live at peace with each other. But sometimes members of a group crowd together so closely that a latecomer to an area cannot find a resting place and jabs nearby bodies with its tusks. If a large bull is doing the jabbing, the sleepy animals are likely to move aside. Otherwise there may be a nasty battle.

On an adult the neck, shoulders, and upper chest have a thick covering of lumpy skin which serves as a shield against the stabbing thrusts of rivals.

It was once believed that walrus tusks were used for digging clams from the muddy bottom of the sea. Today there is some doubt about this. It is thought that the tusks merely glide along on top of the mud while the animal feels for clams with its muzzle, and that the tusks serve only for fighting.

As long ago as the Ice Age people were the enemies of elephants. In those ancient times elephant ancestors (called mammoths) were so huge they would make modern elephants seem small by comparison. But even so, cave men found ways to kill them.

Fossil bones show that the mammoths were driven, or tricked, so that they were made to tumble into large pitfalls. After they were trapped in this way, the hunters bombarded them with rocks, sharpened to make them more deadly. Neither the mammoths' tusks nor their great strength could help them then.

The tusks of our modern elephants, like the animals themselves, are smaller than those of the mammoths—but still they are enormous teeth. Records show that some are well over ten feet long. Both females and males have them, those of the males being larger and heavier.

Today, as with their ancestors, elephants have no enemies except people, and tusks are rarely used as weapons except when bulls are fighting each other.

Elephants are herd animals, and at times there is fierce competition for leadership among the males. Duels are fought with great clashing of tusks. And when an old leader is defeated, he is usually driven from the herd by the victor.

Bulls may also fight for the possession of a female. Young males often duel with one another, but this is more in play than serious challenging.

These young elephants are sparring with their tusks more in play than anger. But when they are full-grown adults, bull elephants fight fiercely to win the leadership of a herd. (Arthur Markowitz from National Audubon Society)

It has been said by close observers that although the bulls fight each other for power, the real leadership of a herd often is carried on by a wise old female elephant.

The tusks of wart hogs, wild pigs, and peccaries are much smaller than those of elephants—but then the animals themselves are smaller. While some of the tusks seem useless because they are overdeveloped, some are vicious weapons.

Probably the best of all for fighting are those of the African wart hog. They grow near the end of the animal's broad snout, curving outward, upward, and inward. They may be only a few inches long, or more than two feet, but long or short, they serve as weapons and to uproot the ground.

Wart hogs live on grasses, roots, and tubers, having no need to kill other animals for food. However, they must defend themselves against flesh eaters and, if cornered, they fight ferociously.

When two boars fight each other, many battle scars are sure to result.

Peccaries are not true pigs, but they are like them in many ways. Like them they eat vegetable matter, including roots and tubers. They have short, strong tusks which are put to good use both for digging and fighting. If a lone hunter, such as a coyote, makes the mistake of getting mixed up with a group of peccaries, it is likely to be ripped and torn to pieces.

When two male peccaries fight, the sounds of cracking teeth mingle with angry squealing.

The tusks of a wart hog give it courage to stand its ground against flesh-eating animals much larger than itself. (H. W. Kitchen from National Audubon Society)

6. *The Deadly Quills*

A PORCUPINE DOES NOT look very frightening. It shows no horns or tusks or other weapon that might alarm an opponent. However, it may injure or even kill—most painfully—any animal foolish enough to attack it.

Its secret weapons are its quills. They are small, being about an inch and a half long on the North American porcupines. And when one of them is not in danger, the quills lie smoothly, apparently no more dangerous than a coat of hair.

Actually the quills *are* hairs. They grow out of the skin in the same way that ordinary hairs grow, from head to tip of tail. But they are hollow, and are tapered at each end with a horny surface.

The danger from a quill lies in the numerous tiny barbs that grow from the shaft. They begin just below the shaft's sharp tip and increase in numbers for a short distance down toward the root.

When quills become imbedded in flesh, the barbs work their way in deeper and deeper. Because of the barbs, it

is impossible for a victim to pull out the quills, and often they continue to move inward until they penetrate a vital organ. Since they may cause inflammation and infection, death may result.

Porcupines have also the more usual kind of hair. It is thick and blackish, and in winter it grows long enough to hide the quills.

People often make the mistake of believing that a porcupine can shoot its quills like so many tiny darts. In reality the animal must come within touching distance of its opponent, and strike out with its tail to hit the aggressor.

Quills are attached loosely to the porcupine's skin, and they come out easily; perhaps a dozen or more will be released with a single blow. If these hit the mouth, nose, or paw of any animal, it is likely to retreat in a hurry. If they strike other parts of the body, the immediate pain is not so great. Nevertheless they can prove fatal if, later, they cause internal harm.

The porcupine is not a fighter by nature. It moves about slowly, on the ground and in trees, feeding on green vegetation most of the summer and on tree bark in winter. But its flesh is tempting food for such mammals as bobcats and foxes. Sometimes they succeed in their attack by flipping the porcupine over with a quick motion of a strong front paw and going after the soft underside.

A porcupine does not begin to fight until it turns its back to an opponent, such as a fox. Then it lashes out with its tail, and its quills may quickly pierce the nose, mouth, eyes, or paws of the animal. (*Ed Cesar from National Audubon Society*)

A porcupine usually prevents this from happening, however. Overtaken by a determined enemy, it bristles—stiffening its spines until they stand up, much as the hair of a frightened cat does, and the quills on the tail rattle, making a sound much like that produced by a rattlesnake. Then quickly turning its back to the attacker, it lashes out with its clublike, quilled tail.

Despite its many thousands of quills, a porcupine can be patted without any discomfort if it is relaxed and unafraid.

In Africa, Asia, and parts of Europe there are "crested" porcupines, so called because of very long guills on the nape of the neck and along the back.

A crested type looks more dangerous than those of North America because of these quills. It also acts and sounds more frightening because it is quick-moving and can speedily counterattack while growling fiercely.

However, it has weaknesses. Its skin is very thin and delicate and the head is unprotected. Altogether it is more likely to be overcome by an opponent than its North American relative.

The most deadly foe of North American porcupines is the small, sleek fisher—a relative of the weasel. It is very clever about getting past the bristling tail, and then quickly sinking its sharp teeth into the porcupine's throat or underside.

Not only that, when a fisher swallows, or is pierced by, quills, it does not suffer from them the way other animals do. It has a natural immunity, and the wounds do not become infected.

7. *Underground Fighting, Odors, and Armor*

Weasels and their close relatives are among the most vicious of all mammals. Though small, they are tremendously strong and quick. They can kill animals many times their size, and often they apparently kill for the sake of killing—not necessarily in self-defense or because they need food.

The black-footed ferret, which is a member of the weasel family, is especially noted for its underground hunting. Prairie dogs are its favorite prey.

The ferrets do kill these animals above ground, but in the open they are not so easy to catch; within the burrows, a catch is easier.

Although the prairie dogs have little chance of escape, a ferret is not always successful. Sometimes one goes into a burrow only to back out rather quickly, bearing wounds from the prairie dog's sharp teeth.

Prairie dogs are constantly alert for enemies. On occasion, when one is returning to its burrow, its sense of

smell reports that a ferret has gone down there and is lying in wait. Quickly, then, the prairie dog scratches earth over the openings to the burrow, burying the enemy alive.

After a while the ferret pushes its way out to freedom. But in the meantime, the prairie dog has found a hiding place.

It seems strange that the fierce, fast-moving ferrets belong to the same family group as the slow, easygoing skunk. But they do; and one feature they have in common is the power to give out a bad-smelling odor. However, it is only the skunk whose scent is powerful and sickening enough to keep possible enemies at a safe distance.

A skunk's ammunition is a musk that is held in two large glands at the base of the tail. When the animal senses danger, it contracts muscles that surround the glands and, as a result, the musk shoots out in a fine spray. In a brisk wind the smell can carry for half a mile or more.

Only one "scent gun" may be shot at a time, or both may go off together. Though the first blast is the strongest, enough musk remains for more to follow. After the glands have been emptied, it takes a while before they are again filled with "ammunition."

An underground den makes a good home for a prairie dog in some ways. But when this animal is followed into its burrow by a ferret, it can be trapped and killed there. Prairie dogs bark or give shrill whistles when alert to danger. This one is barking at the entrance to its den. (*Harry Engels from National Audubon Society*)

A skunk is not quick to use its valuable weapon. When it sees something dangerous approaching, it is likely to growl and stamp a foot angrily. If this does not frighten away the opponent, the tail goes up—a sign of "ready, aim, fire!"

When a well-directed shot blasts into the face of an enemy, this unhappy creature howls in pain. Besides the dreadful smell, the spray stings the eyes, perhaps causing blindness for a while. Most animals need only one experience with a skunk to stay well away ever after.

Another member of the weasel family is the wolverine. It is the largest member; adults weigh up to fifty-five pounds. Like the smaller weasels, it is quite fearless.

Wolverines have a need to be good hunters and fighters because they live where the ground is snow-covered through long winters and food is scarce.

A wolverine stakes out a large home range, choosing a territory of many acres. Its ownership rights are advertised by leaving its strong musk scent at various positions. If it comes on a wolf or mountain lion eating from a kill, it may end the feast by a sudden attack. Often the other animal gives up rather than being torn by the wolverine's sharp teeth and claws.

Wolverines can bound along with all four feet hitting the ground at the same time, but even so, their speed is no match for such long-legged mammals as the caribou. Therefore their strategy is to climb into a tree or on a high rock and wait there. When an unsuspecting victim comes close enough, the wolverine leaps and quickly sinks its teeth into the neck of the enemy.

Food usually is scarce in the cold areas where wolverines live, and they must be good hunters and fighters to survive. This wolverine is challenging two coyotes over the rights to food discovered in the snow. (Ed Cesar from National Audubon Society)

For many years the wolverine had a reputation for being able, and willing, to eat more than its own weight of food at one time—whenever there was that much food to be found. There is no truth to this. Once it has caught its own prey, or discovered a dead animal, it will eat heartily. Then it sprays what it cannot finish with its musk, and the foul smell discourages other hungry ones from coming close.

An unusual means of defense is for an animal to grow its own armor. The armadillo is one mammal that does this.

Armadillos are creatures of the New World, and when Spanish explorers first saw them, they gave them the name which, to them, meant "little armored thing." (In Spanish *armado* is "a man with a coat of armor.")

Some armadillos are larger than others; the biggest grows to a length of five feet. But all species have the armor-like shell.

The shell is made up of many bony plates, welded together and covered with scales. It protects the animal's back, while another shield does the same for the head. Even strong teeth cannot bite through such a defense.

In spite of its rigid covering, an armadillo can move rather easily and quickly because flexible bands across the middle of the shell act as hinges. The animal can bend, twist, and roll into a ball—and it does so as a defensive act because the underparts have no protective cover.

An armadillo's defense against flesh-eating animals lies in its armor-like shell, which can resist even strong, sharp teeth. (R. F. Head from National Audubon Society)

Enemies such as the fox try to turn an armadillo over to reach the soft flesh. And sometimes if the armored one has rolled itself into a tight ball, the attacker pushes it into water, causing it to unwind.

Armadillos have no means of fighting back. Their teeth are placed far back in the jaws, and they are blunt rather than sharp. The giant species does have a long, curved claw on each front foot. But the claw's real importance is for digging and tearing apart termite nests to get at the armadillo's favorite food.

The pangolin of Asia and Africa is another mammal with armor. This armor is more than defensive, however; the large overlapping scales of which it is made are pointed and can be used as weapons.

Pangolins can move more freely than armadillos. Like them, they roll into a ball to protect the soft underparts. But if an attack continues, the pangolin raises its scales and hisses savagely. The next move is to lash its tail, driving its knife-like scales into its tormentor. The scales on the body also can give severe wounds.

Added to a pangolin's defenses are glands from which a bad-smelling spray can be shot. It is something like a skunk odor, but not so strong.

Pangolins can hold their own very well against small flesh-eating animals. Large enemies, such as tigers, find them helpless, however.

8. *Snake Fighters*

MAMMALS FIGHT other mammals. And at times they also fight other kinds of animals. Such as snakes!

Often it is the mammals that are the attackers because they find the reptiles make good food. But there are also fearful encounters between giant snakes and mammals which end with the mammals being swallowed by the reptiles.

Boa constrictors and pythons kill their victims by constriction, although their first move may be to strike the prey with their sharp teeth. Immediately, then, the snake throws coils of its body forward so that they are wrapped around the mammal. Next, two or three of the coils are tightened so that the victim is prevented from breathing, and soon its heart stops beating. When it no longer moves, the snake loosens its coils—and eats.

Such encounters are frequent with small pythons and boas, with their prey being no larger than mice and rats. However, because of their small size, these animals do not attract special attention. The spectacular fights are

between something like a twenty-foot-long python and a leopard. In Africa and Asia giant pythons feed on such mammals as antelope and large deer.

Poisonous snakes present a different challenge to their opponents. The snake may be small, but if it can inject a deadly venom into its victim, size hardly matters.

Scientists think that the rattles developed by rattlesnakes came about as a means of giving these reptiles protection against large animals. Since prehistoric times, in the western part of North America, there were many hoofed mammals, including rhinoceroses and horses. These creatures could easily trample snakes to death, even though poisonous species could cause much suffering to any animal they managed to bite.

So it seems that over many thousands of years some snakes had important changes take place in their tails. The skin at the tip became thickened and, from the time of birth, the young ones had a structure on which "rattles" could build—rattles formed by some of the old skin whenever a snake shed its outer covering. Afterward when a rattler's tail was vibrated, a sound resulted that served notice to any large mammal to be careful where it was stepping. The mammals learned about the discomfort of poison that resulted from contact with this "rattling" animal.

Mammals and snakes often are opponents in fights for survival. Some mammals eat snakes; some snakes eat mammals. Snakes use poison or constriction in their attacks. Mammals, such as this bobcat, depend on sharp teeth and strategy to win a battle. (*Courtesy of The American Museum of Natural History*)

Nevertheless, some mammals refuse to shrink from combat with rattlesnakes. One of the most successful in battling them is the peccary, a piglike animal that stands no more than two feet high and weighs less than sixty pounds. Yet a peccary fearlessly charges to within a few feet of a rattler, then stops short, waiting for the snake to strike. When it does, the peccary leaps up and, with its back arched and feet pulled together, it comes down on the rattler's head, slashing with its hoofs. Over and over this action is repeated until the snake is destroyed.

The mongoose is especially noted as a killer of poisonous snakes. It deserves its fame, but there have been many exaggerations about its ability.

Snakes are an important food for mongooses, and these mammals certainly do not hesitate to tackle the poisonous species. In Africa they battle the dreaded black mamba, and in Asia cobras often become their victims. However, the mongooses are not completely immune to the reptile poison, and sometimes they are the losers in a fight.

An experienced mongoose works like a boxer, provoking the snake to strike, then dancing out of reach. When the snake drops flat on the ground, fully extended, the mongoose leaps close and tries to sink its teeth in the reptile's neck. This strategy is repeated over and over until a fatal wound quiets the snake, which the mongoose then eats, poison glands and all.

It is thought that a mongoose may develop some degree of immunity from a constant eating of poisonous snakes. Certainly some of these animals can handle the venom better than others.

When snake fights snake, as these two rattlers are doing, the action is called a combat dance. However it is not a "dance," but a real struggle and test of strength. (R. Van Nostrand from National Audubon Society)

There is a belief among native people of some tropical countries that when a mongoose is bitten by a poisonous snake, it immediately hunts in the jungle for a certain root to eat, and this prevents any ill effects. However, this is not proven.

A peculiar struggle of the animal world is carried on among snakes, particularly the rattlers. It involves two males which intertwine the foreparts of their bodies. Then they strain against each other, swaying back and forth. So graceful are their movements, the performance is known as the "combat dance."

However, scientists call this a serious combat—a contest of strength—rather than a dance. When one snake's head is pushed to the ground, and pinned there by a coil of his opponent, the fight is over. The stronger of the two has won the right to certain territory.

9. *Designs for Peace and Protection*

STRANGELY, the most ferocious-looking of animals—the gorilla—is among the most peace-loving of all.

Gorillas lead a quiet life in the African forests. They have no enemies except people with guns, and occasionally a leopard which might dare to attack a young one. Their only food is vegetation so they are not involved in hunting other kinds of animals.

These giant apes live in family groups, made up of males, females, and young. Each group has an especially large and powerful male as a leader.

Once in a while a leader feels threatened by the leader of another family, and this may cause some show of defiance. However, if groups happen to meet, they are likely to be friendly until each goes its own way.

A leader is the decision-maker. His group follows where he leads, to feeding grounds and resting places. He watches out for dangers and when he suspects any, puts on a wonderful show of power.

First he inflates air sacs in his chest and gives a series

of loud hoots. Then, pulling himself up to full height, he slaps his chest, belly, and thighs. The thundering noise that results can be heard a mile away.

When his drumming is finished, he drops on all fours and dashes forward, tearing branches from trees and bushes, yanking vegetation out of the earth, and slapping hard at anything in his path. Suddenly he stops and thumps the ground, then calmly sits down.

This display of strength is designed to frighten away any unwelcome intruder. Usually it does unless, of course, the intruder is a person with a gun. If an enemy does continue to threaten, the leader hurries his companions out of the way, staying behind to roar and perhaps actually to charge.

Within a group good behavior also is managed by threats more than by actions. If two females start screaming at one another, the leader glares at them fiercely, and they quiet down. If young ones disturb the peace by too rough play, an older gorilla can restore order by slapping the ground with a very determined thump.

As a rule, members of a group are quite polite to each other. If a top-ranking individual wants another to step aside, it points this out with a gentle tap of its fingers. This may be followed with a push. However, it then will probably wait until the other gorilla moves of its own free will.

Chimpanzees also can make a forest ring with their hooting and screaming. But this may be done with no anger at all. It seems that when two groups from different areas suddenly meet, great excitement results. The members scream, but there is no fighting.

Sometimes two individuals get into a scrap, but it usually is a short one. After a scuffle, they chase each other, and then the quarrel seems forgotten.

Orangutans are the quietest of the apes. However, when they feel threatened, they make a variety of unpleasant noises, gulping, grunting, barking, and giving tremendously loud burps.

People watching orangs in a natural setting have seen them act like show-offs, as when a male would pose in a tree, feet and hands anchored on the branch below him. Then suddenly his hands would be raised and he would dive forward and down, his feet holding securely so that he could hang upside down.

Orangs spend much of their time among tree branches. If a family is disturbed by some intruder, such as a human, the male will come down toward the ground, looking as frightening as possible, while the female and young ones quietly climb higher.

Orangs also noisily break off branches and hurl them at any possible enemy below.

Gorillas, chimps, and orangs living in forest areas are given considerable protection by the trees. Baboons, however, belong more to the open plains and, although they sleep in the trees, most of their days are spent on the ground.

Sharing this same territory—the plains of Africa—are a number of flesh eaters, including leopards and lions. One baboon, or even several, would be no match for such large and powerful beasts. But these monkeys flourish because of their strong defense pattern.

A group, or "troop," of baboons is well organized, both when on the move or at rest. At the center are the females with their babies and other young ones. Close to them are the strongest males. At the front and rear are the less important males. In this formation the group goes in search of food and waterholes. When an enemy is detected, however, the strong males quickly move out front, and lead the other males in confronting a dangerous prowler.

Baboons live peacefully among other animals that graze on the African plains, even though all eat the same kind of food. Often they benefit from these animals which are able to warn of danger. Giraffes, for example, tower above them, and can see far across the open country. Gazelles have exceptionally keen senses of hearing and smell. Therefore, when animals such as these show alarm, the baboons are immediately put on alert.

A frightening face and the size and strength of a giant combine to make gorillas feared as vicious monsters. Yet in their natural surroundings these great apes are peace-loving and even gentle. (Maurice E. Landre from National Audubon Society)

Among other mammals that band together for safety are musk oxen. The "wild cattle of the far North," when threatened by wolves, form a tight circle. Young calves stand in the center of the circle, which is made up of both males and females whose hooked horns, pointed threateningly, are enough to frighten away even a hungry beast.

In a wolf pack, friendliness and affection does exist. Nevertheless, there is a "pack order," with a leading, or dominant, male in control. Other males have importance in varying degrees, and the "degree" of each one is understood by all members of the pack. Because of this, a sharp look from a dominant animal is enough to bring a lesser wolf into line.

Today there are few bison remaining on the plains of North America, and they do not live under completely natural conditions. But years ago, millions of these animals made a spectacular sight, and many careful accounts were kept of their habits before civilization began to destroy them.

The enormous herds were composed of small bands of perhaps fifty to a hundred members. Each had a leader, usually one of the older females. As they moved about their range, she walked at the head of the group while the bulls guarded the other cows and young ones.

When a cow was about to give birth, she moved a short distance away from her herd. After the calf was

Wolves are fierce hunters in the quest for food, but among themselves there is a social order that helps keeps peace in a family or pack. These young ones are having only a playful scuffle. (Joe Van Wormer from National Audubon Society)

born, it would be several days before the little one could walk firmly, and the mother stayed with it constantly. If wolves appeared, she covered the calf with her body and loudly bellowed a plea for rescue.

Quickly the bulls of the group would appear and make a ring around the mother and baby. Those forming the circle all faced outward, lowering their heads so that their horns pointed at the wolves, ready for action.

At all times bulls kept a close watch over any calf that had become separated from its mother.

10. *The One-sided Contest*

OVER AND OVER we see that in the animal kingdom "kill or be killed" is the way of life. However, there are interesting facts to notice:

Animals kill other animals because they are meat eaters and must do so to obtain food. Mostly they kill only when they, or their families, are hungry. Otherwise they are as peaceable as the plant eaters.

Fights between male members of the same species are often only a measuring of strength. If the combat is for the possession of territory or the winning of a female, as soon as superior strength is shown by one, the struggle ends. (Accidents may change this, as when deer antlers become entangled.)

Certain mammals are said to be particularly savage. The cape buffalo of Africa is one that has such a reputation. But the fact is, the "savage" behavior occurs only when a buffalo is being hunted, particularly if it has been wounded. Otherwise the great animal is quiet and shy, and seeks only grass and seeds as food.

Another reason for animals having an undeserved reputation is that some (not all) "wildlife films" give false impressions of their dispositions.

For instance, in order to show a very dramatic scene, unnatural fights are promoted. Someone drives animals that usually stay within a certain territory into the territory of others of the same species. Those whose homesites are invaded respond by attacking. Or a female may be taken from her mate and tied near another male. Fighting between the two males is then a certainty.

People who see many such programs may be left with the feeling that animals all are hostile to each other, and are constantly "at war."

Actually, wars such as are fought by people—with one group trying to destroy another because of hatred—are unknown to animals. But people *are* known to them; unfortunately they are known as enemies. And against such "opponents" as guns, traps, poisons, and bulldozers that destroy wilderness areas, there is little to be done.

The contest between people and wild creatures is very one-sided.

A "dominant" wolf shows its high rank in the pack by resting its head on the neck of one of less importance. (Douglas H. Pimlott Photo)

Index

Page numbers in italics refer to photographs..

Index

DOROTHY SHUTTLESWORTH has been exploring nature since she started working for The American Museum of Natural History in New York at the age of seventeen. After several years there on the staff of *Natural History,* she became the first editor of a similar magazine for young people—*Junior Natural History,* a position which she held for twelve years. Mrs. Shuttlesworth lives in East Orange, New Jersey, with her husband, a retired high school principal. Her book *Litter—the Ugly Enemy,* was cited as an outstanding book by Keep America Beautiful, Inc. *Clean Air, Sparkling Water, Litter—the Ugly Enemy,* and *To Find a Dinosaur* were all selected by the National Science Teachers Association and The Children's Book Council joint committee as Outstanding Science Books for Children in 1973. Her *Disappearing Energy* was given similar recognition in 1974.